The Cartography of Farmers' Wives

Photography from 1915-1976

By Colleen Weber Borst
Grace Shaw Weber & Emma Heckner Shaw

The Cartography of Farmers' Wives

Photography from 1915-1976

By Colleen Weber Borst
Grace Shaw Weber & Emma Heckner Shaw

Introduction

By Colleen Weber Borst

In 1912 my great-grandmother, Emma Heckner Shaw, moved from Salt Lake City, Utah, with her husband and one-year old daughter, to the prairies of Alberta in hopes of making a life for themselves. She brought with her an upright piano and a Pocket Folding Kodak camera to keep her company on the flat dry land, and as her husband worked the land, as they moved from farm to farm, or found themselves working odd jobs to make ends meet when the land was fickle and unbearing, my great-grandmother continued to play music and take pictures of her surroundings.

Years later, in the mid 20th century, my grandmother, Grace Shaw Weber, became the wife of a farmer. She photographed the changes to the land around her, and later became a professional journalist and press photographer in the Wenatchee Valley. Her husbands' family was one of the first immigrant families to settle eastern Washington and dry farm the land, and later were instrumental in the Columbia Basin Irrigation Project and the building of the Grand Coulee Dam, which allowed the vast arid land of Eastern Washington to become viable irrigated farm land.

As the land around my grandmother changed, photography was changing as well. While the wheat, vegetables and irises that my family planted grew up from the earth and dotted the horizon with explosions of colour, my grandmother was able to preserve these small bursts on film as colour photography became more common place and accessible.

Now these photographic memories belong to me. Some are family photographs, gatherings of people I will never know, whose faces have slipped out of the collective memory of my family and are anonymously burned in silver on film. But some of these images, mainly the majority presented in this small collection, are contemplations of the land.

By showing a small fraction of the photographs I have been given by the women in my family, I hope to present a quiet meditation of what used to be. The hopes and struggles that went into the stories behind these photographs are largely irrelevant to those outside my immediate family, but I do hope that the sense of history and of time start to creep under your skin when you view these photographs, and I hope they spark a reflective dialogue about your own history, as well as a dialogue about our collective history with the land.

May, 2012

P.S. I still have my great-grandmothers Pocket Folding Kodak camera. While it is a little worn out with time, it still works beautifully.

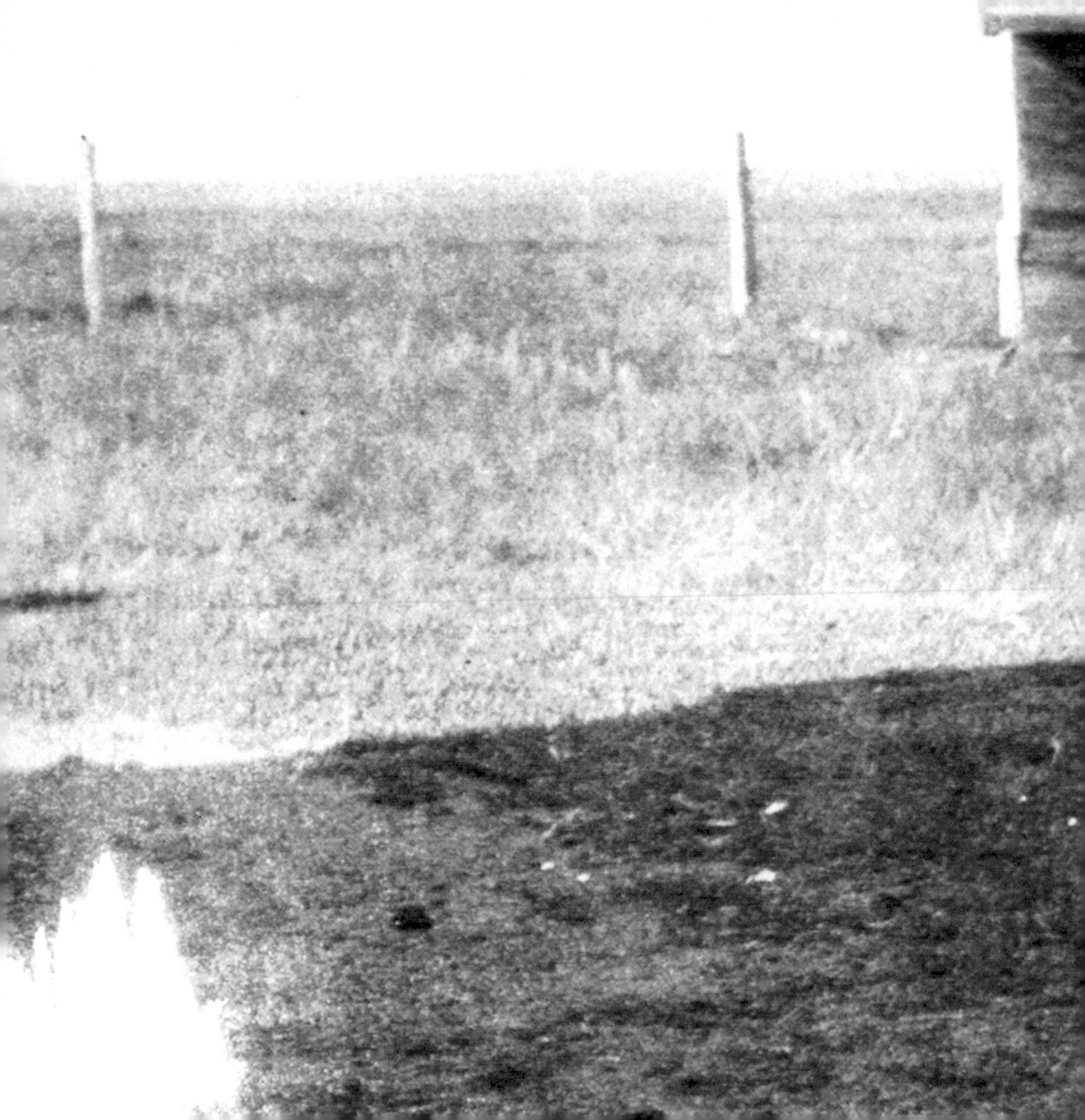

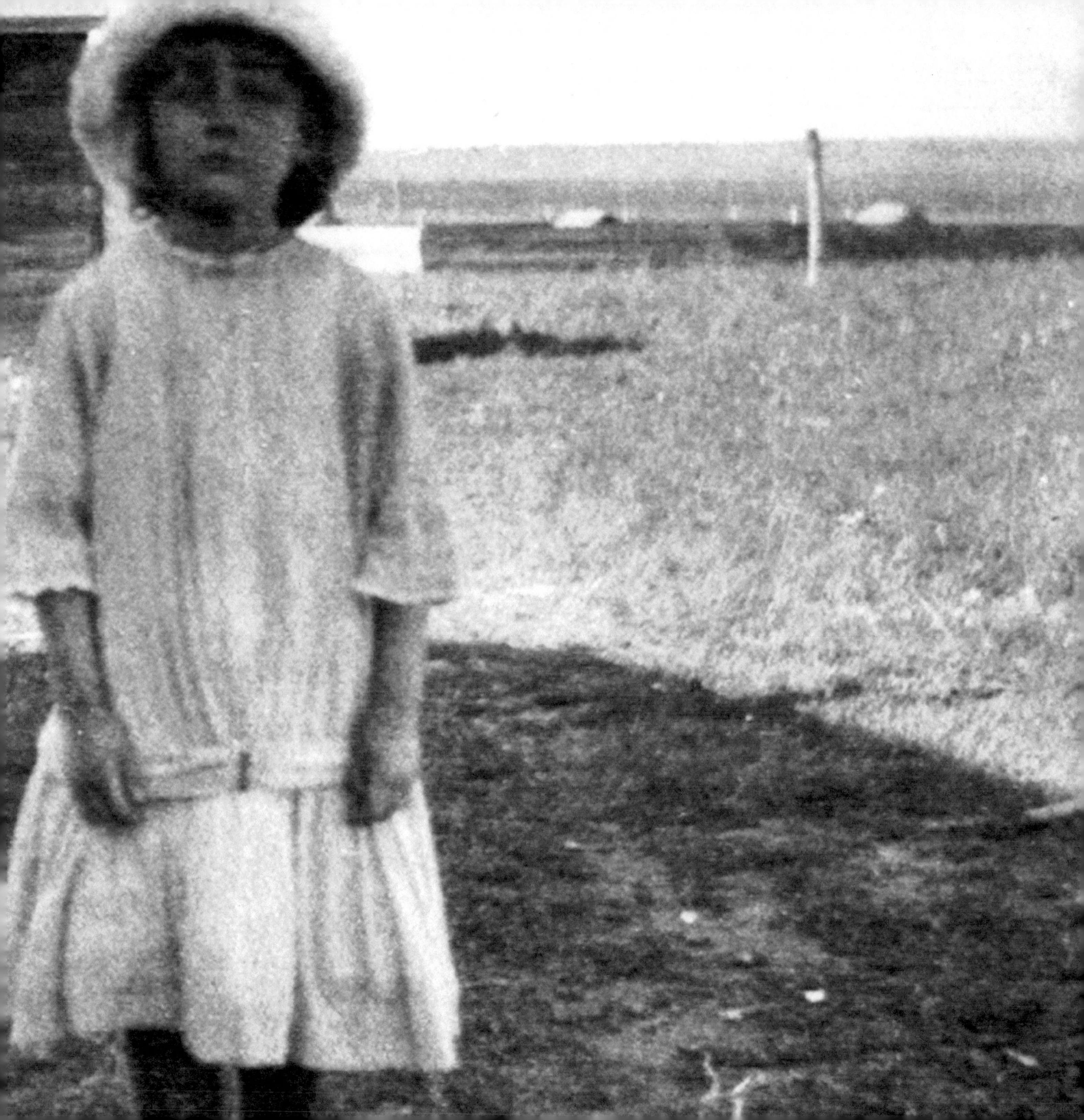

▸ Three Hills, Alberta, about 1915, Emma Heckner Shaw
Three Hills, Alberta, about 1916, Emma Heckner Shaw

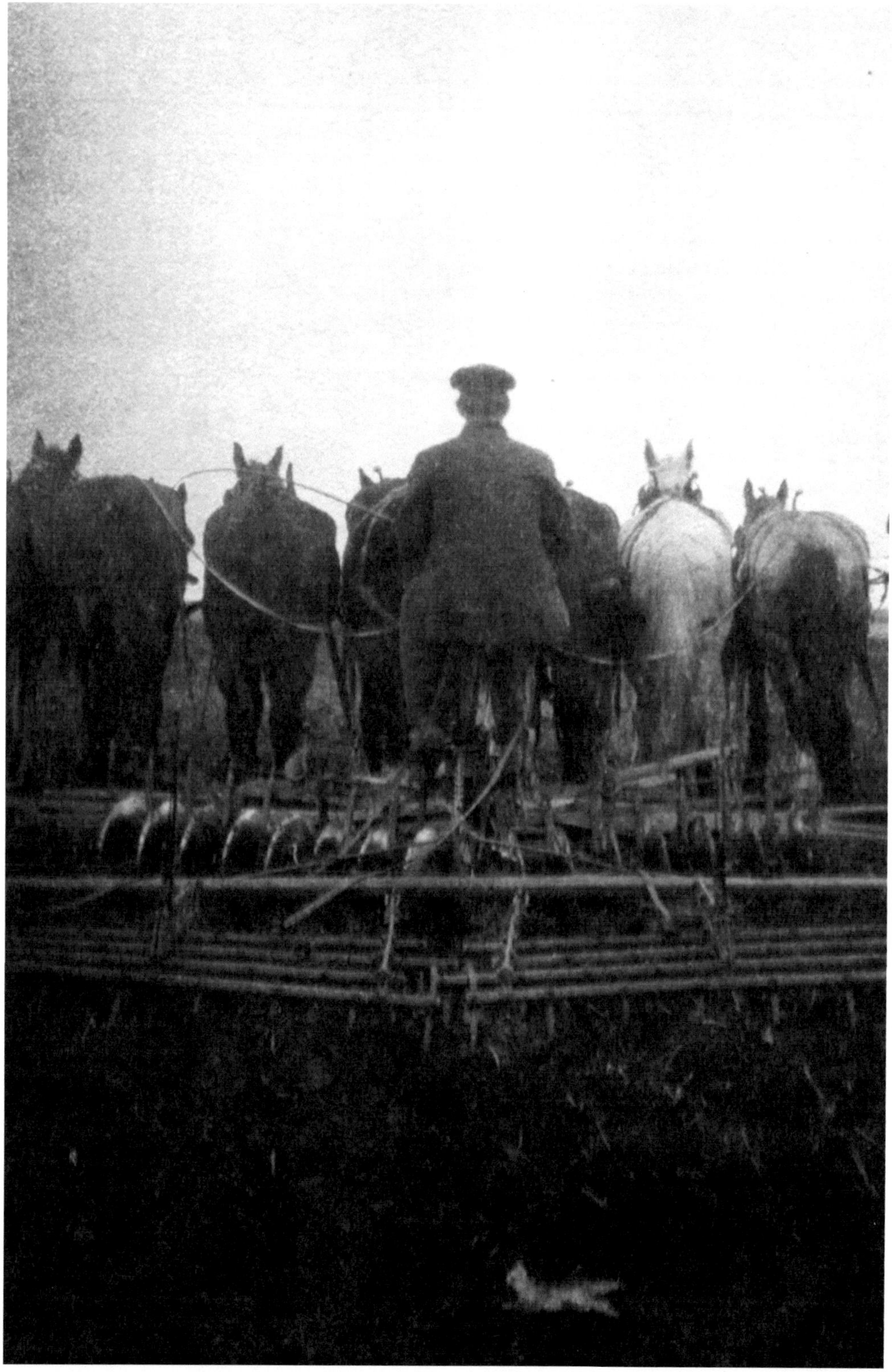

Alberta, about 1917, Emma Heckner Shaw

▶ Banff & Lake Louise, Alberta, about 1918, Emma Heckner Shaw

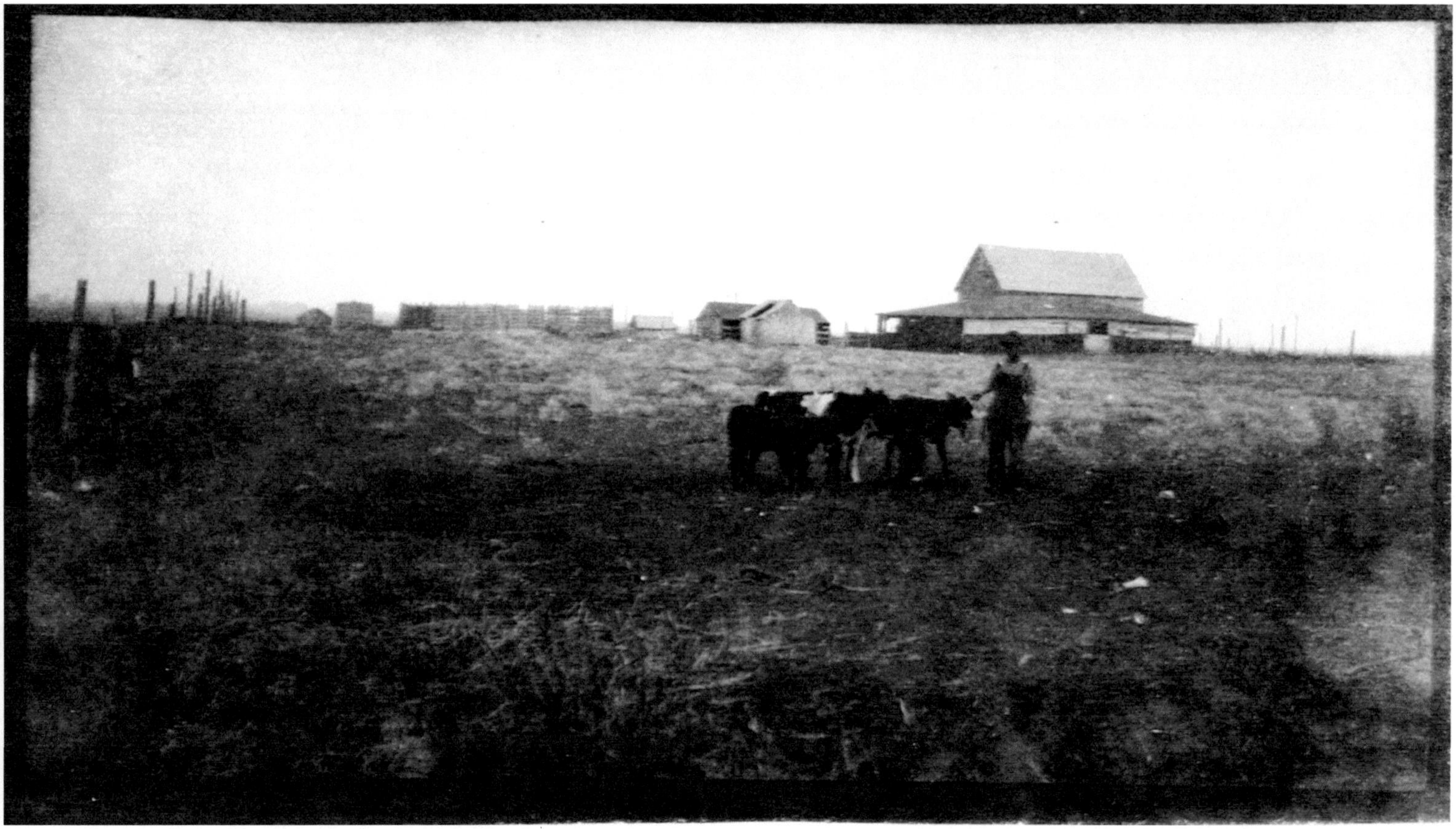

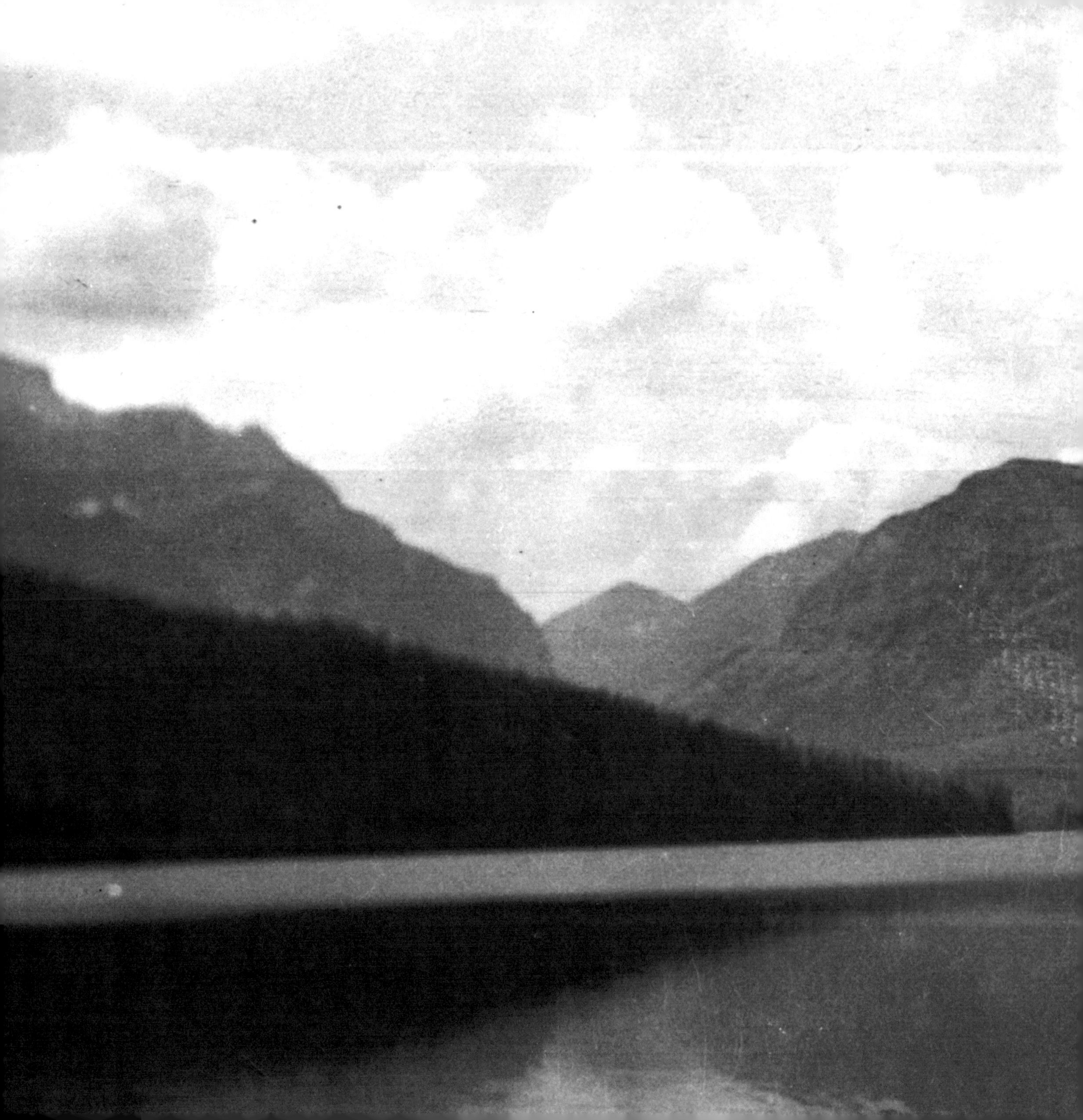

Alberta, about 1918, Emma Heckner Shaw

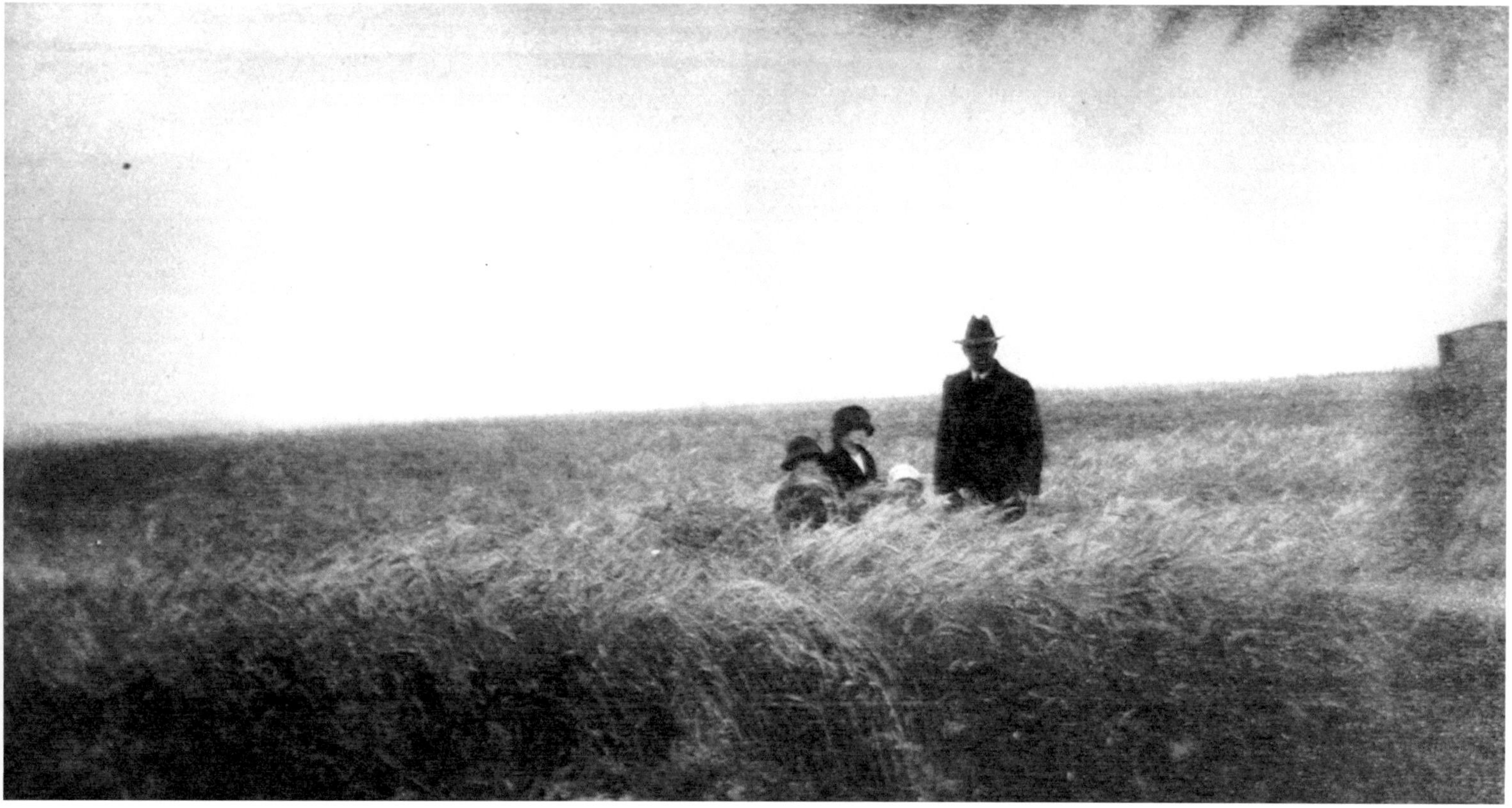

Alberta, about 1919, Emma Heckner Shaw

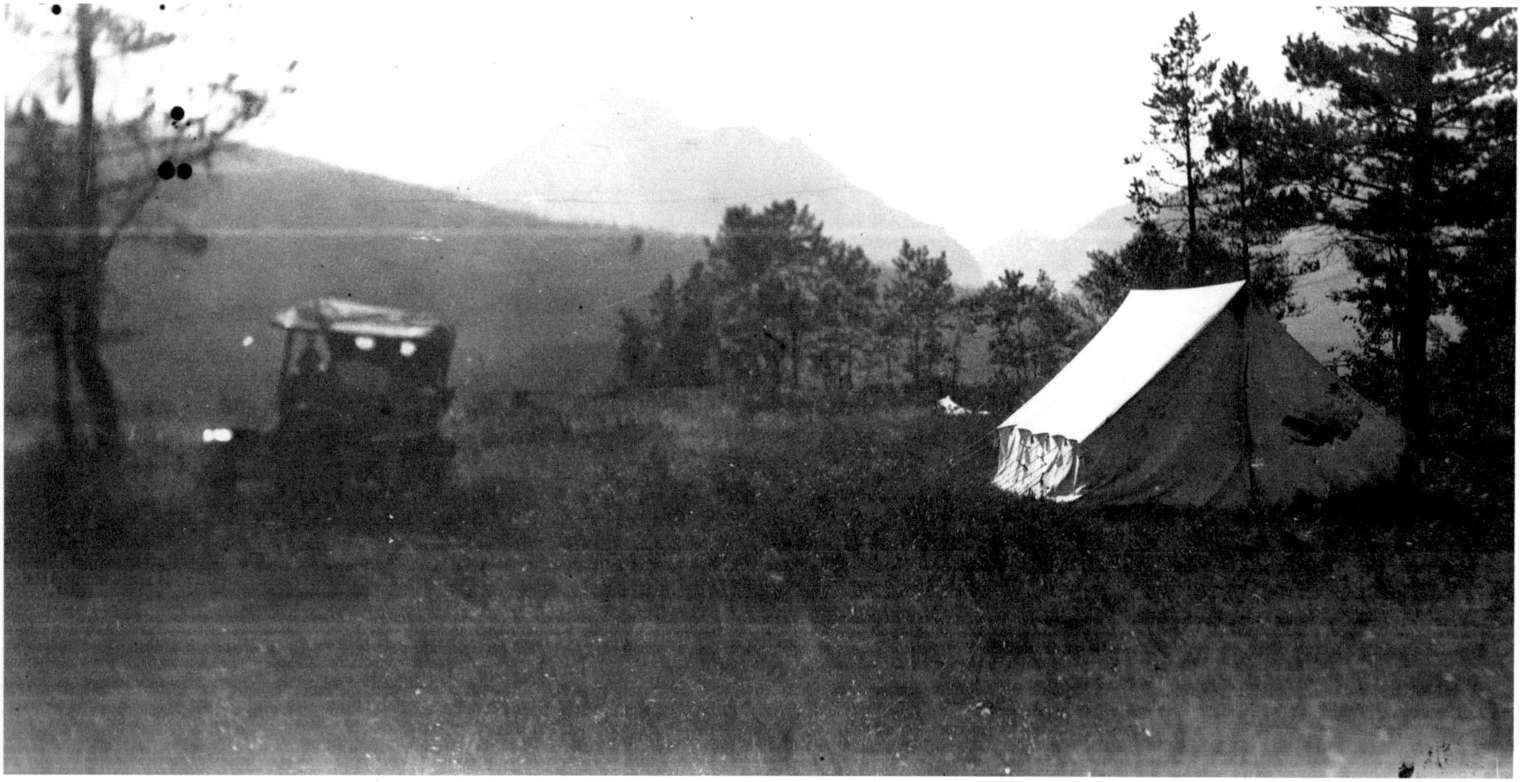

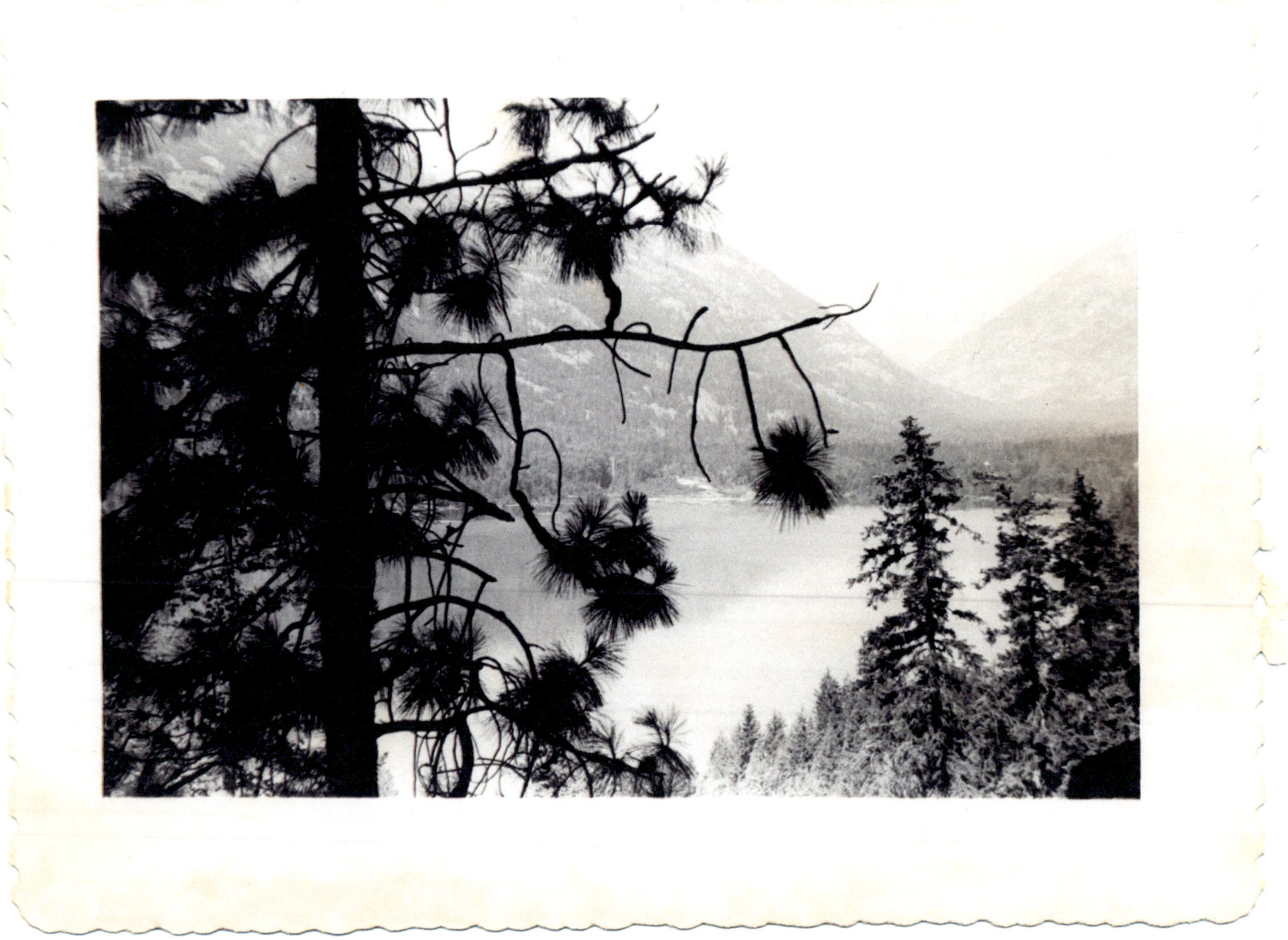

Lake Chelan, late 1940's, Grace Shaw Weber

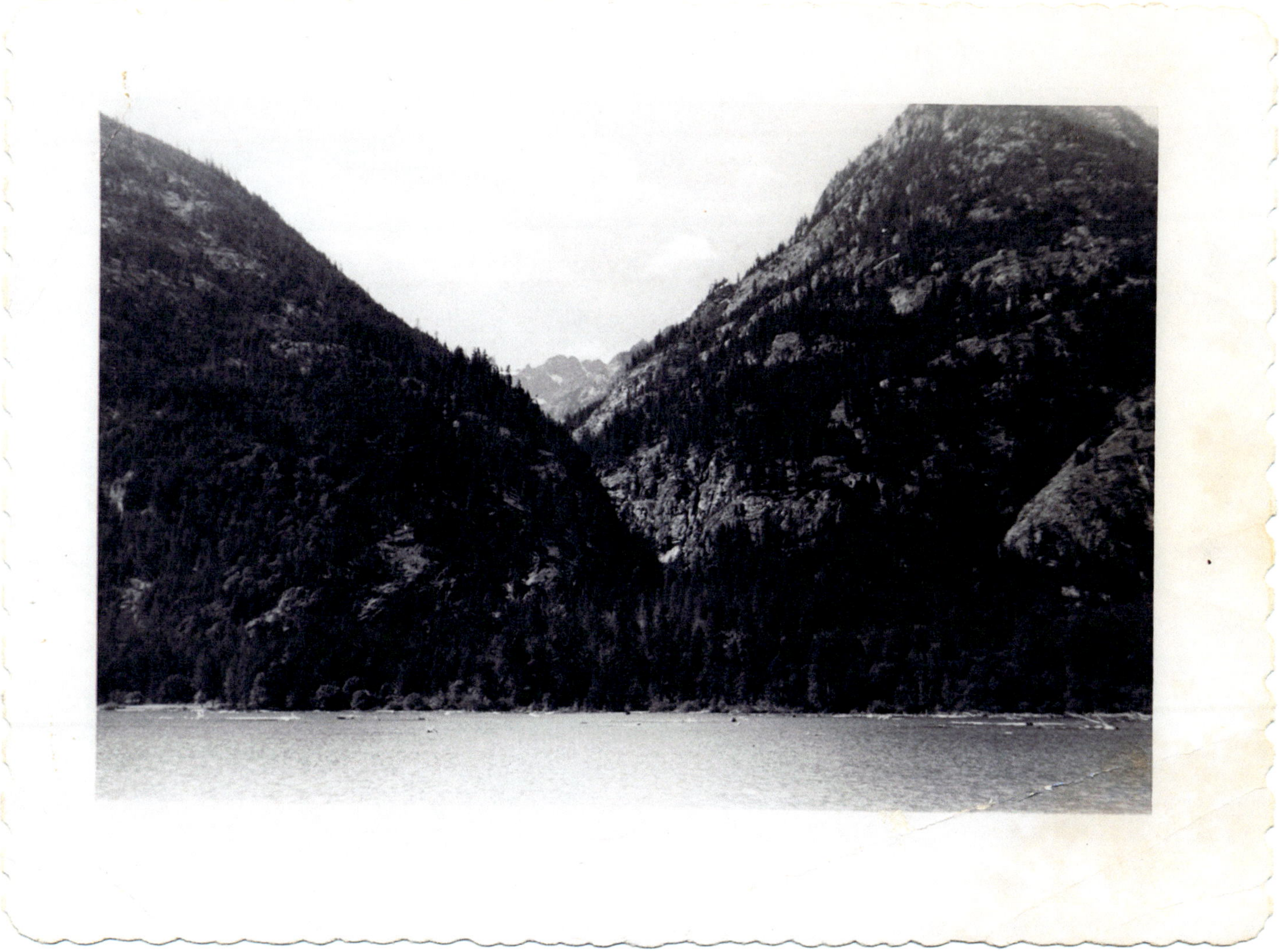

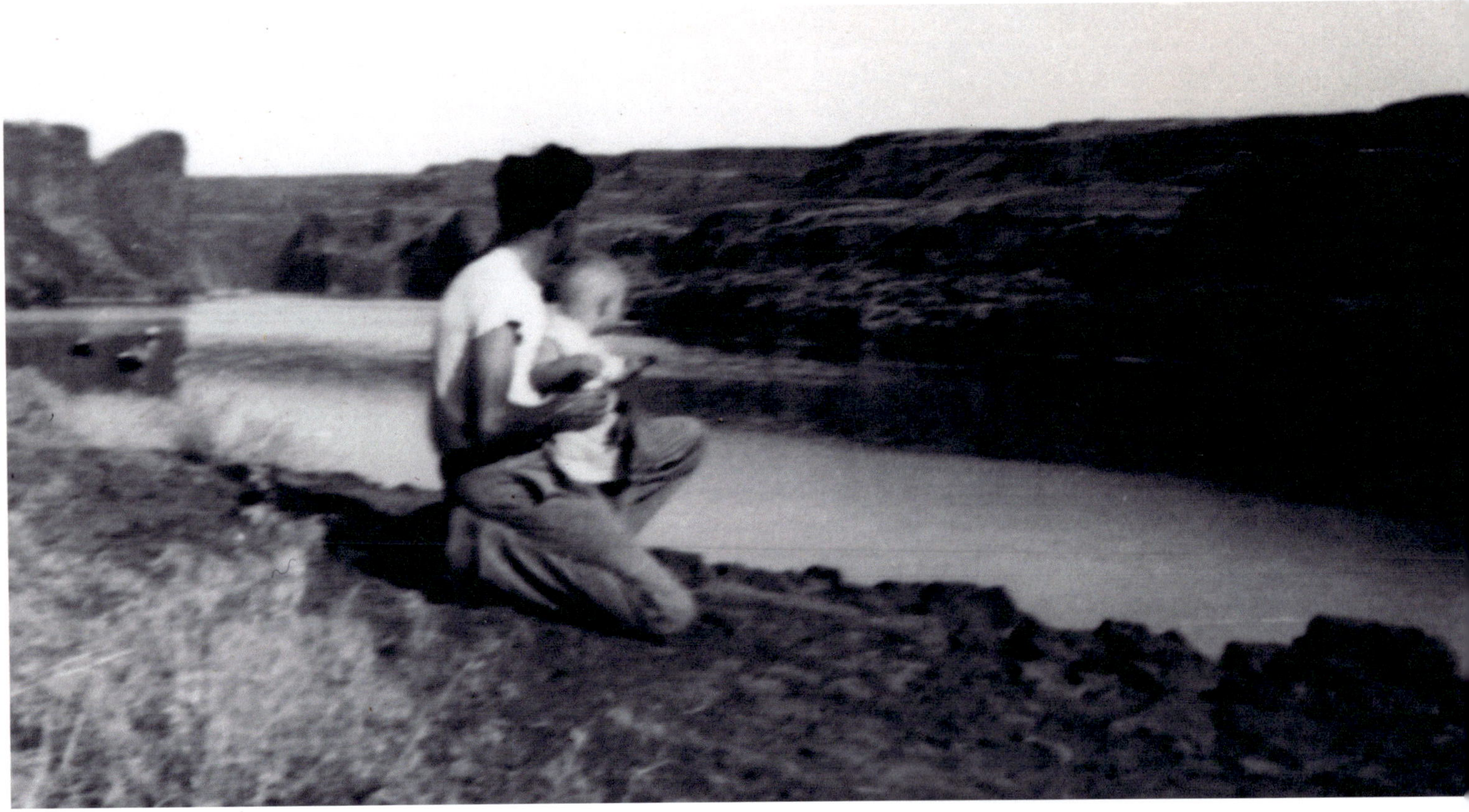

Mid Columbia River, 1948, Grace Shaw Weber

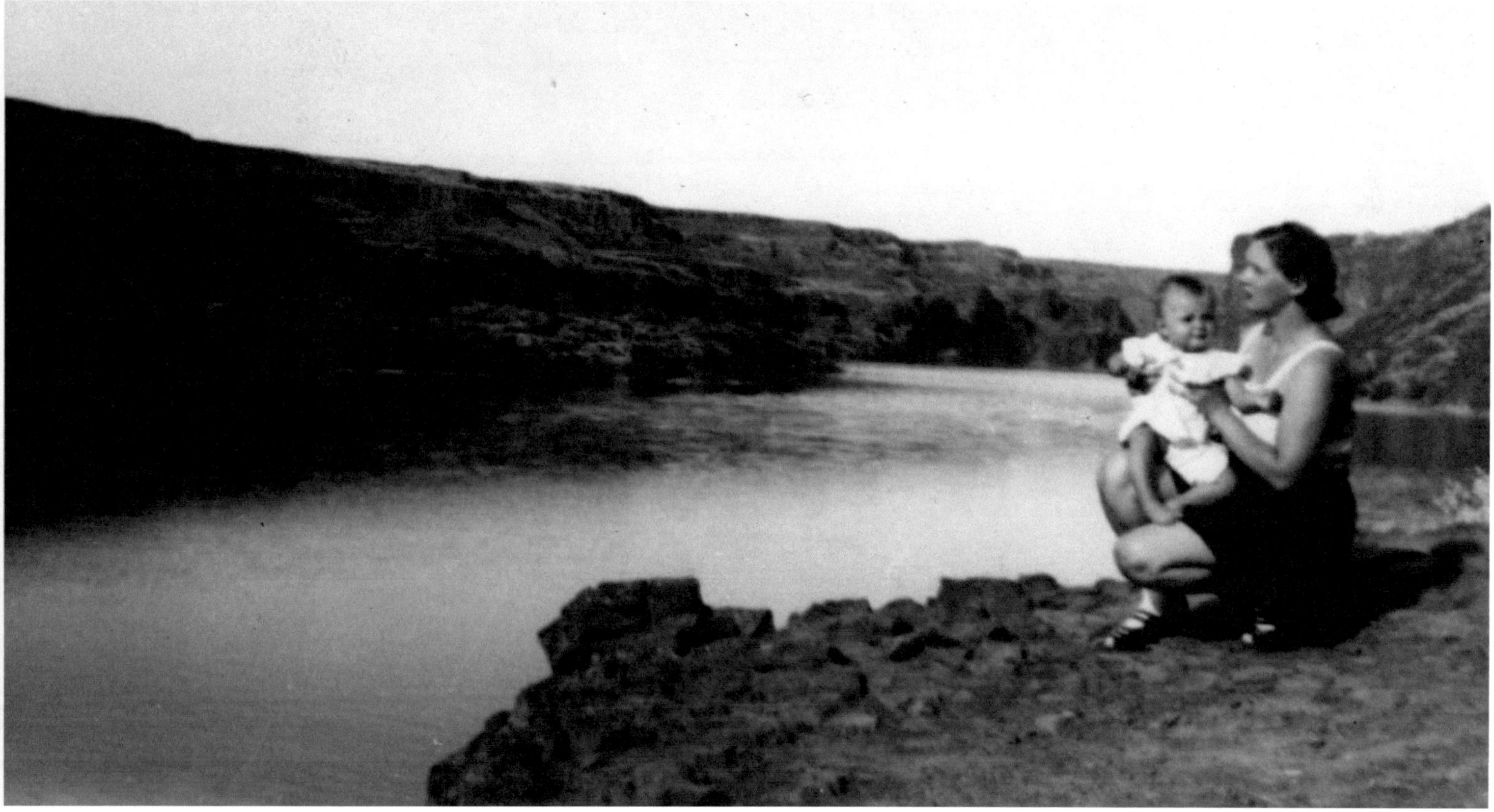

Quincy, 1957, Grace Shaw Weber

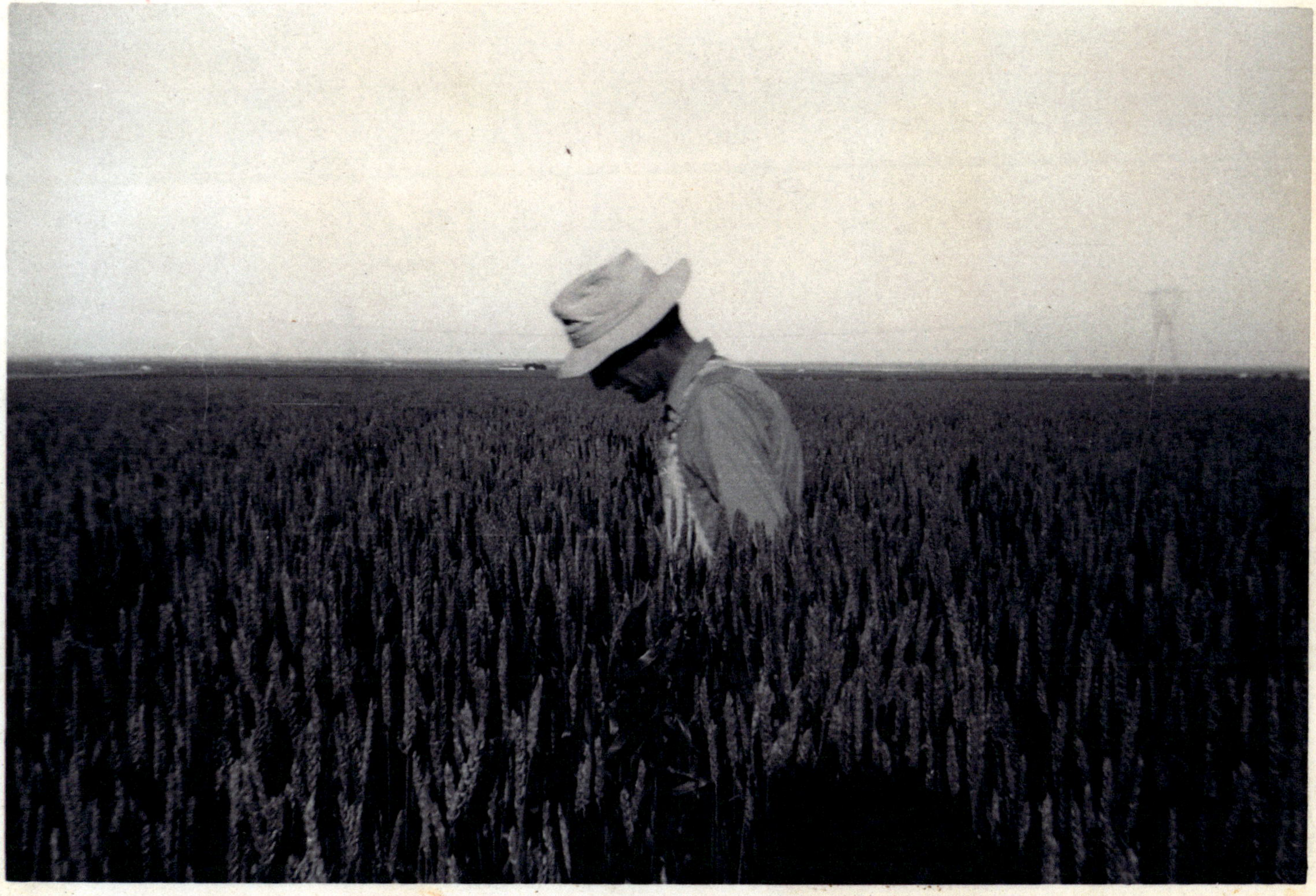

Quincy, 1954, Grace Shaw Weber

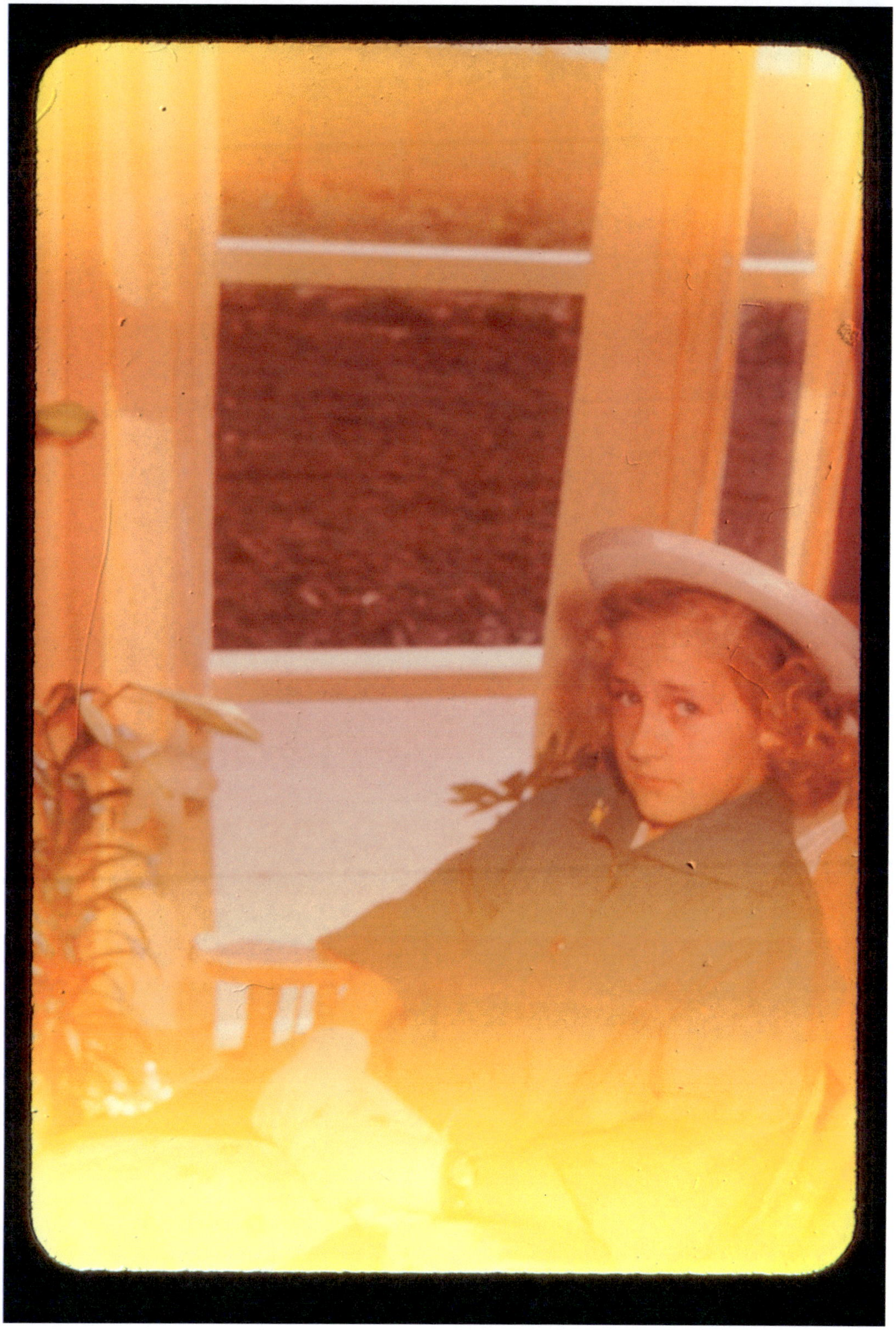

Unknown, 1970's, Grace Shaw Weber

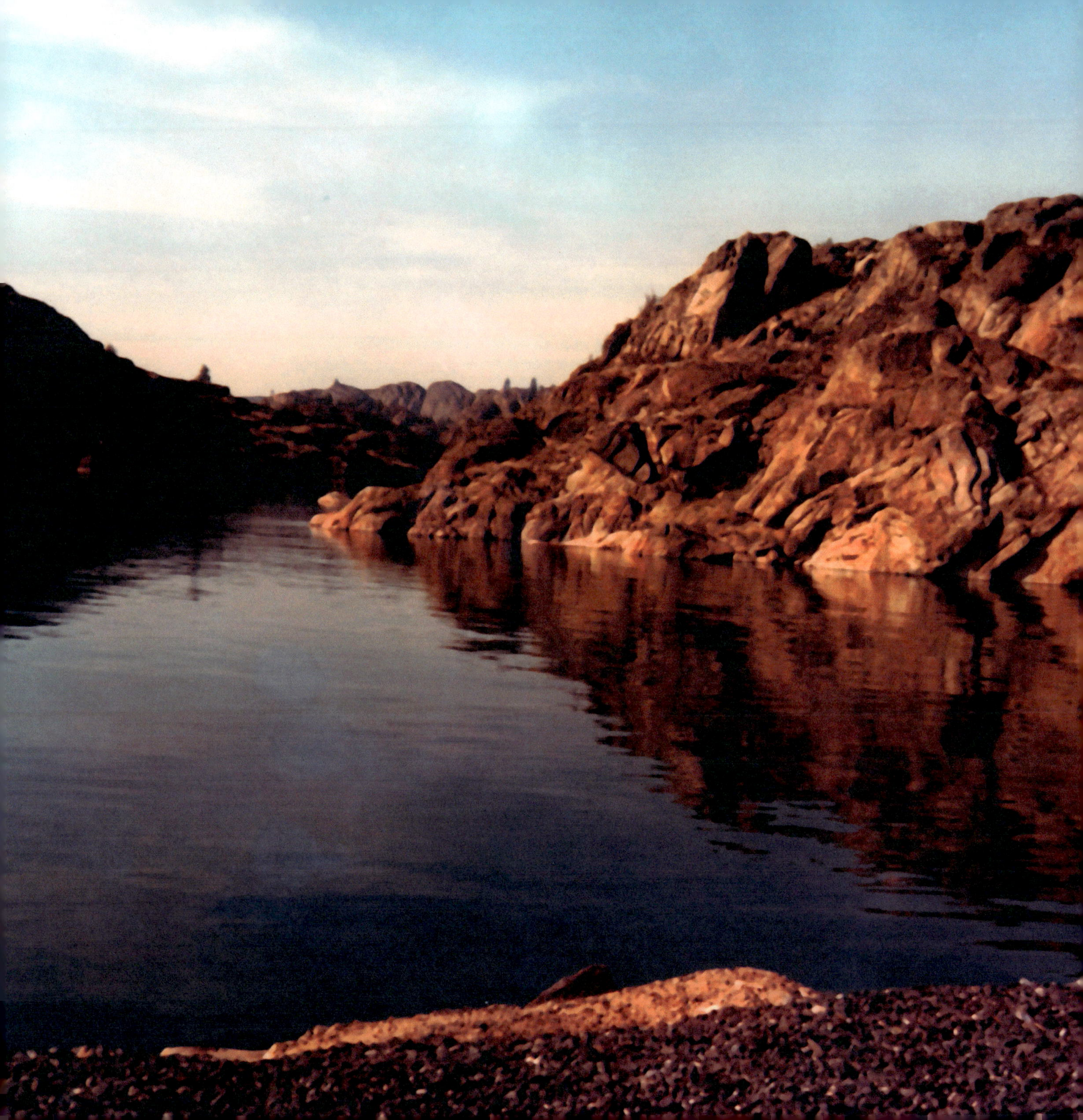

Quincy, 1958, Grace Shaw Weber

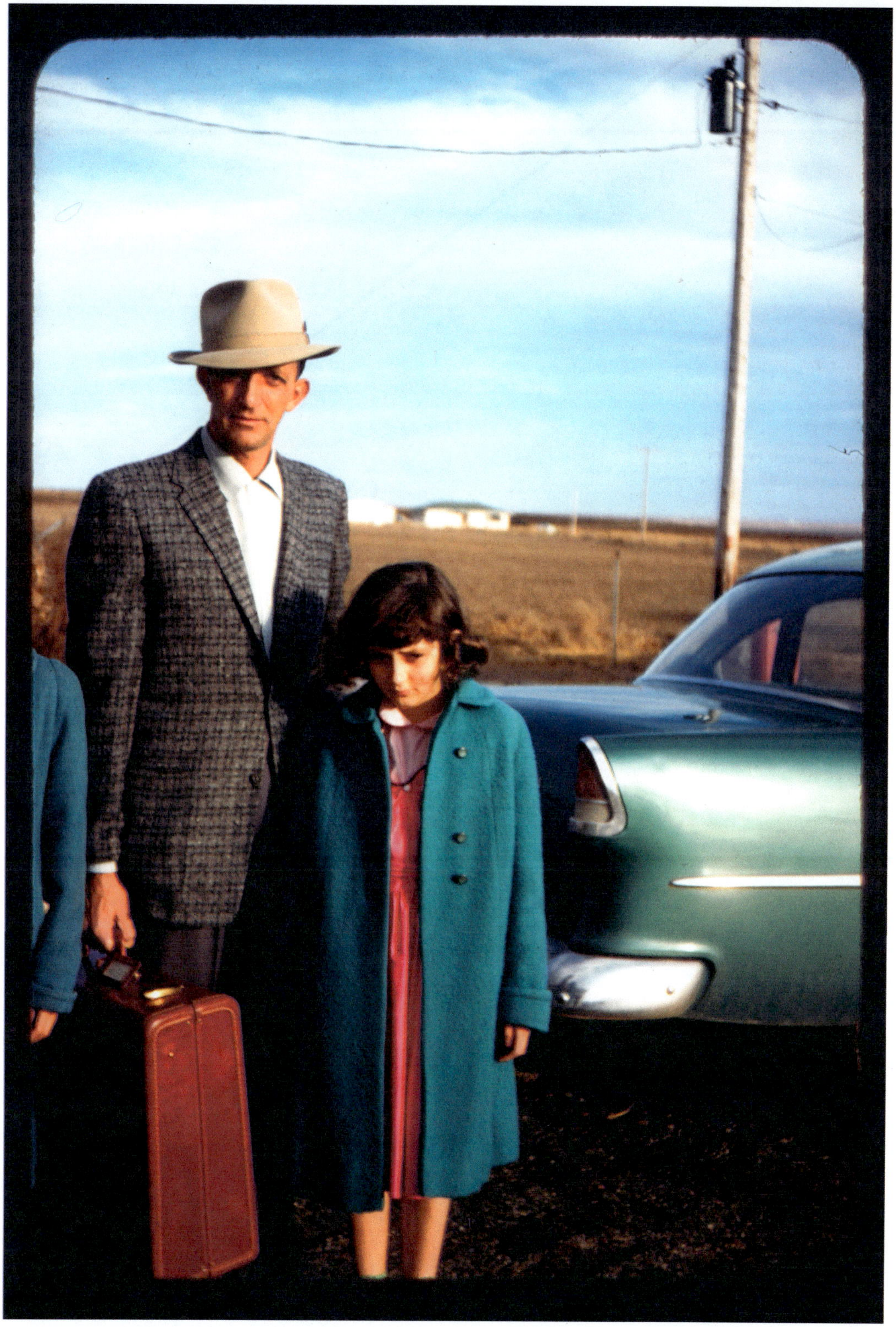

Columbia Cresent Bar, 1960s, Grace Shaw Weber

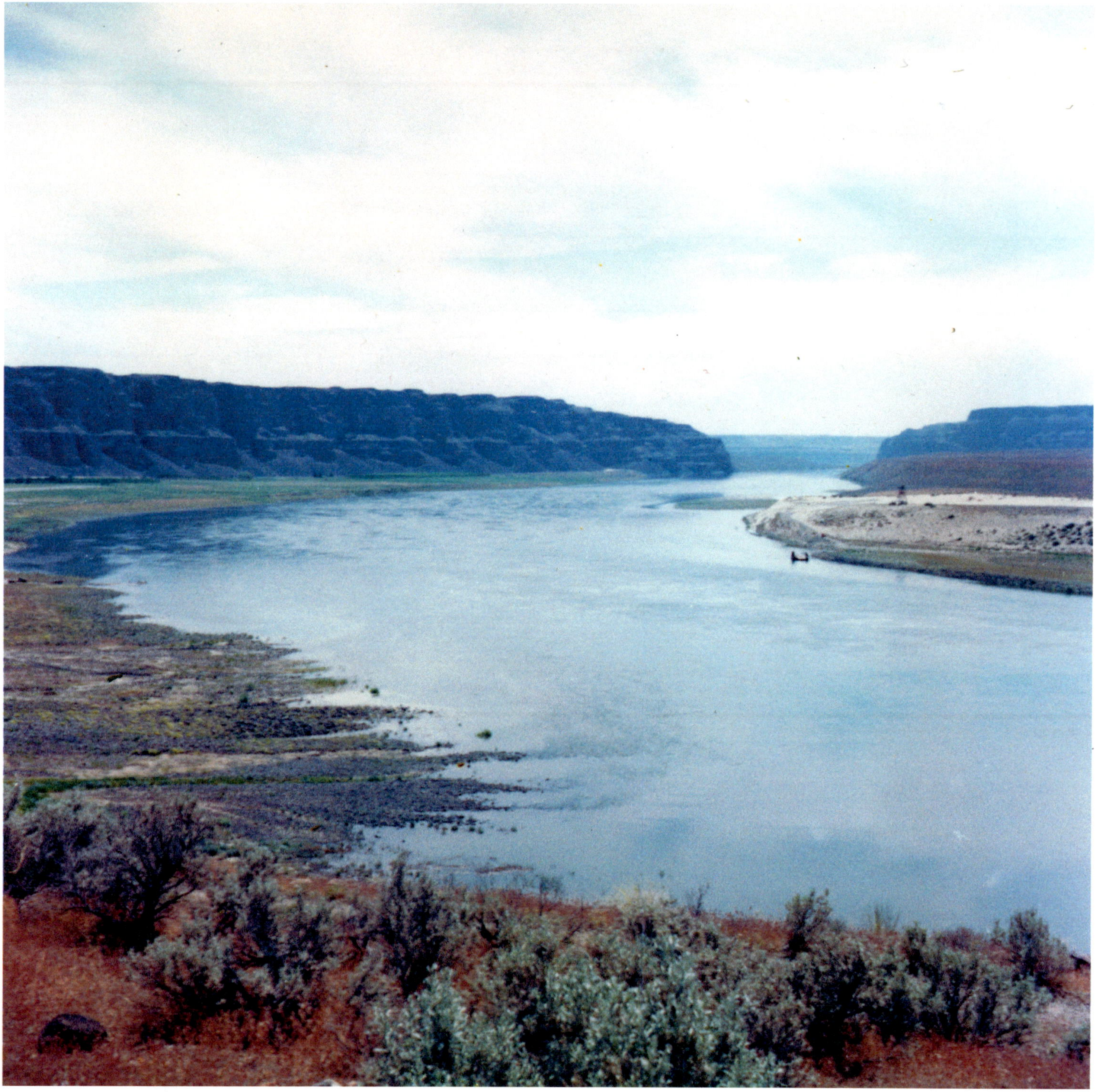

Burien, about 1958, Grace Shaw Weber

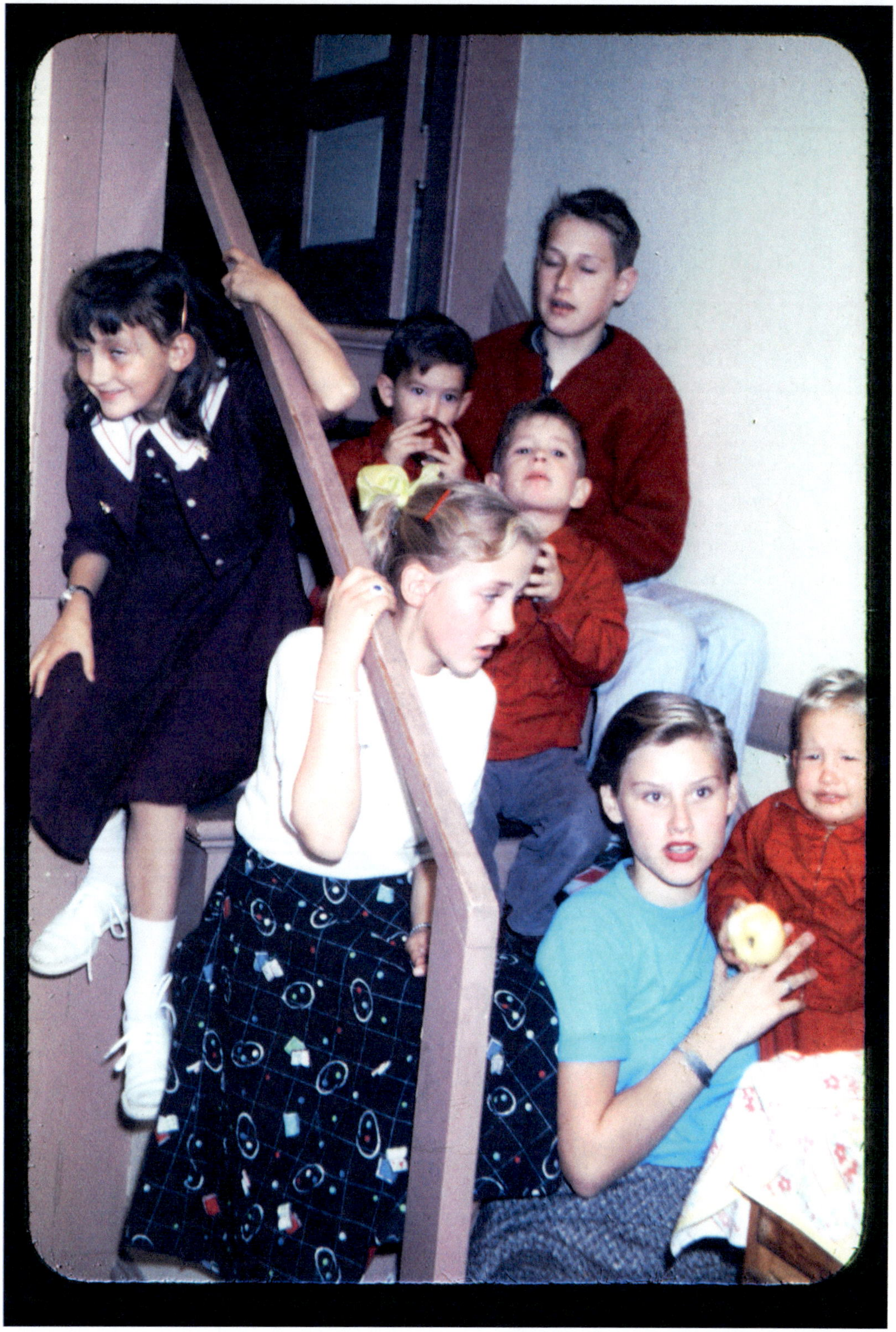

Unknown, 1970's, Grace Shaw Weber

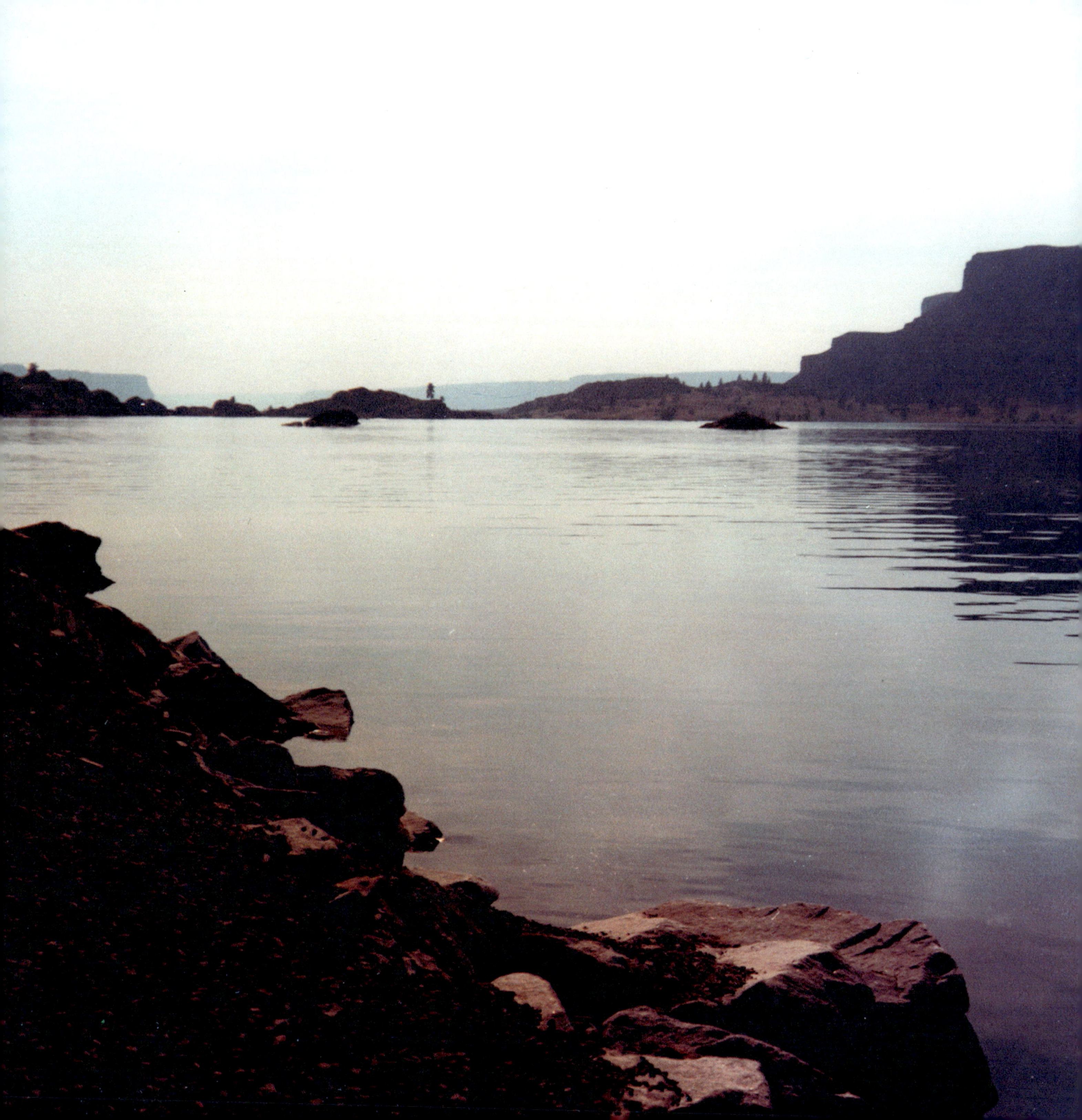

Quincy, about 1960, Grace Shaw Weber

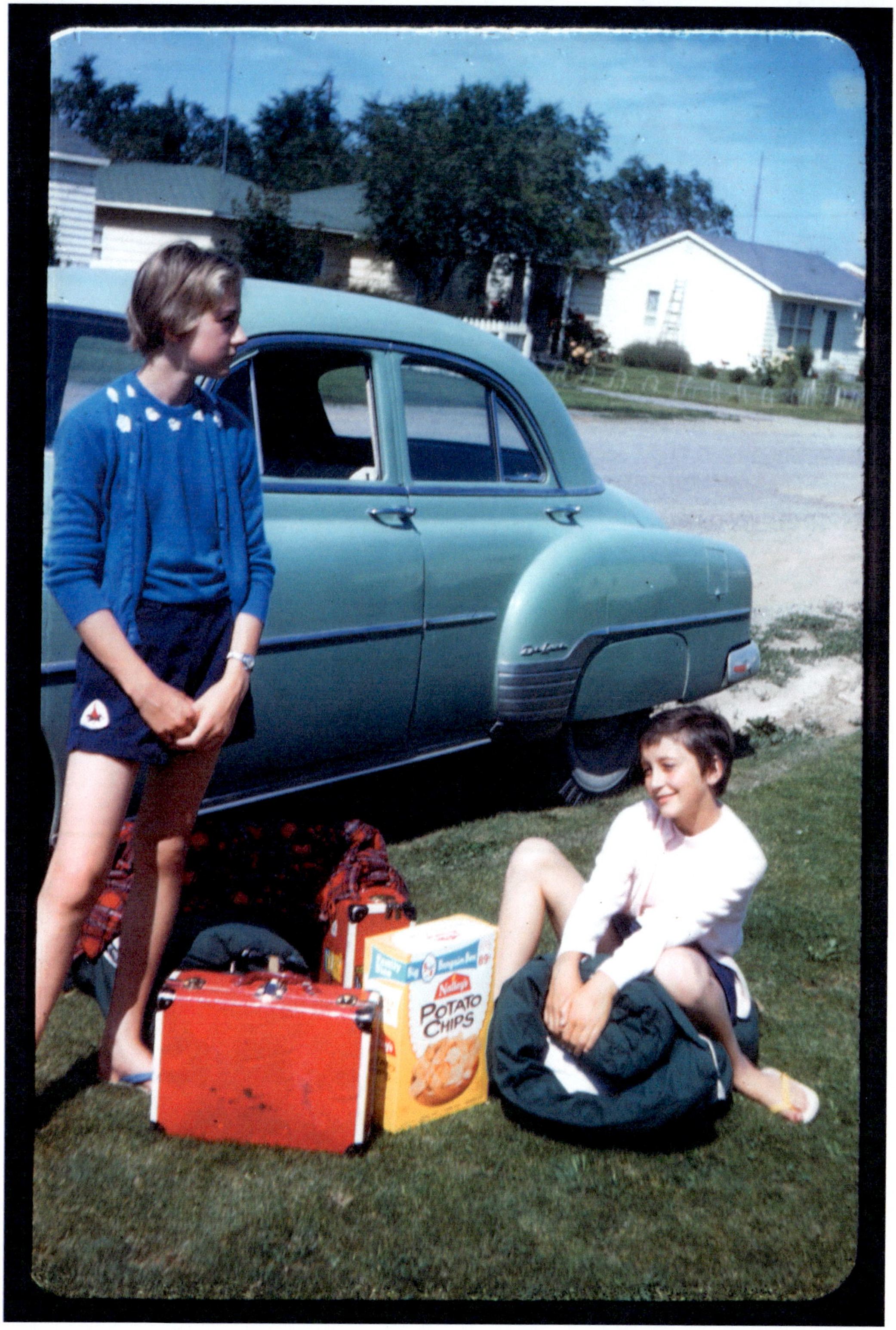
Nalley's
POTATO CHIPS

Kittitas Valley, about 1974, Grace Shaw Weber
▸ Leavenworth, 1970's, Grace Shaw Weber

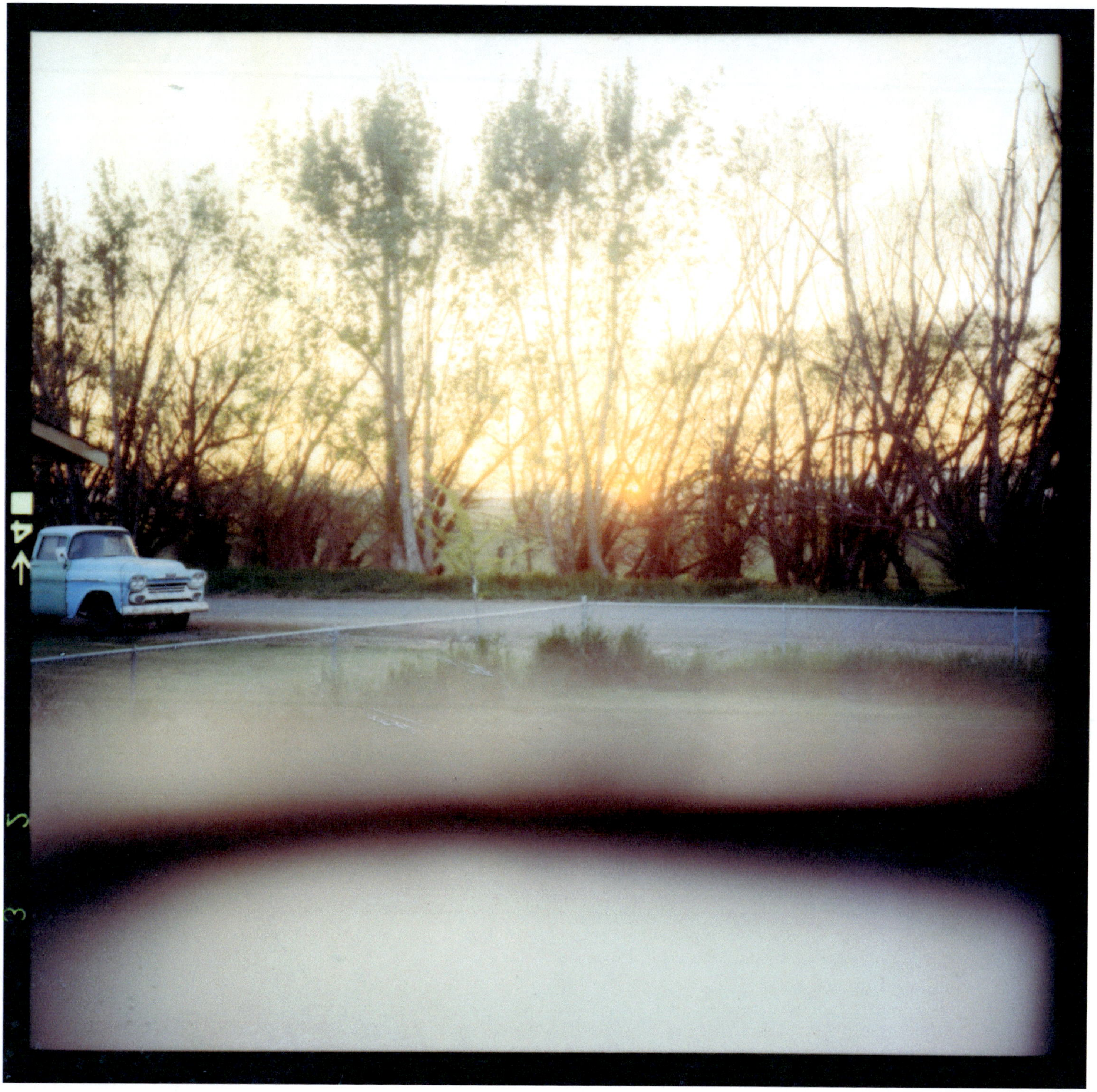

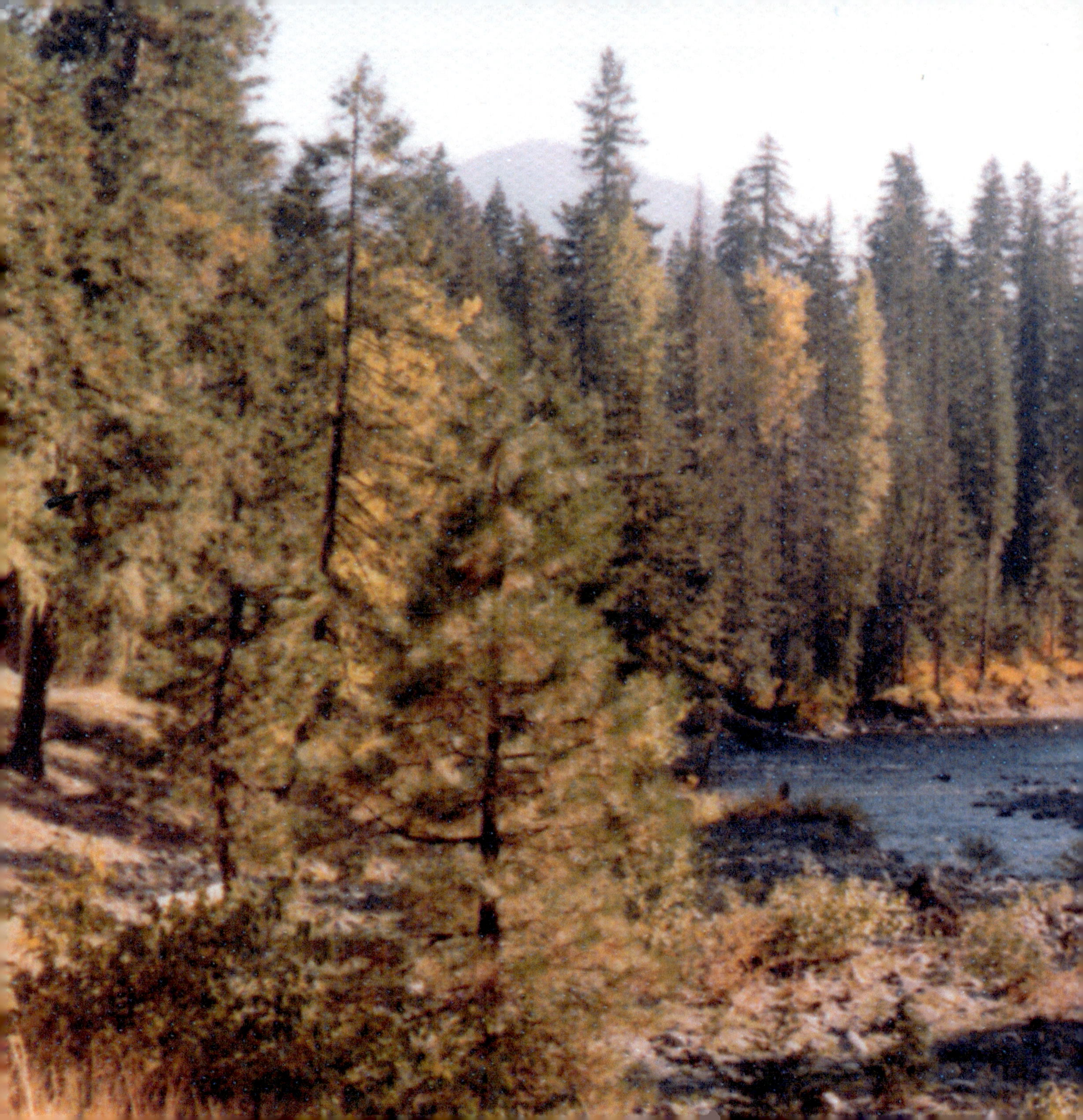

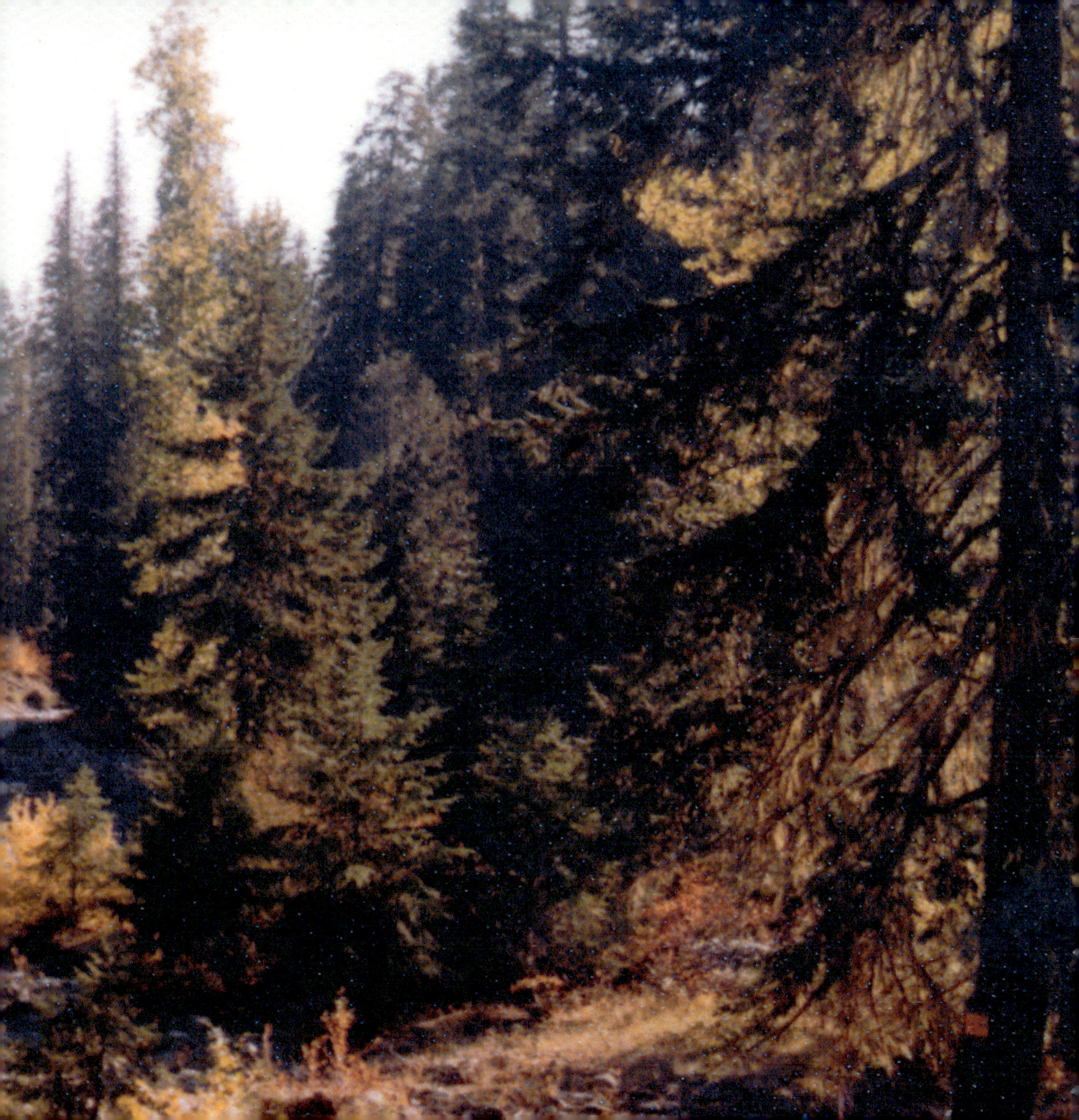

Leavenworth, 1970's, Grace Shaw Weber

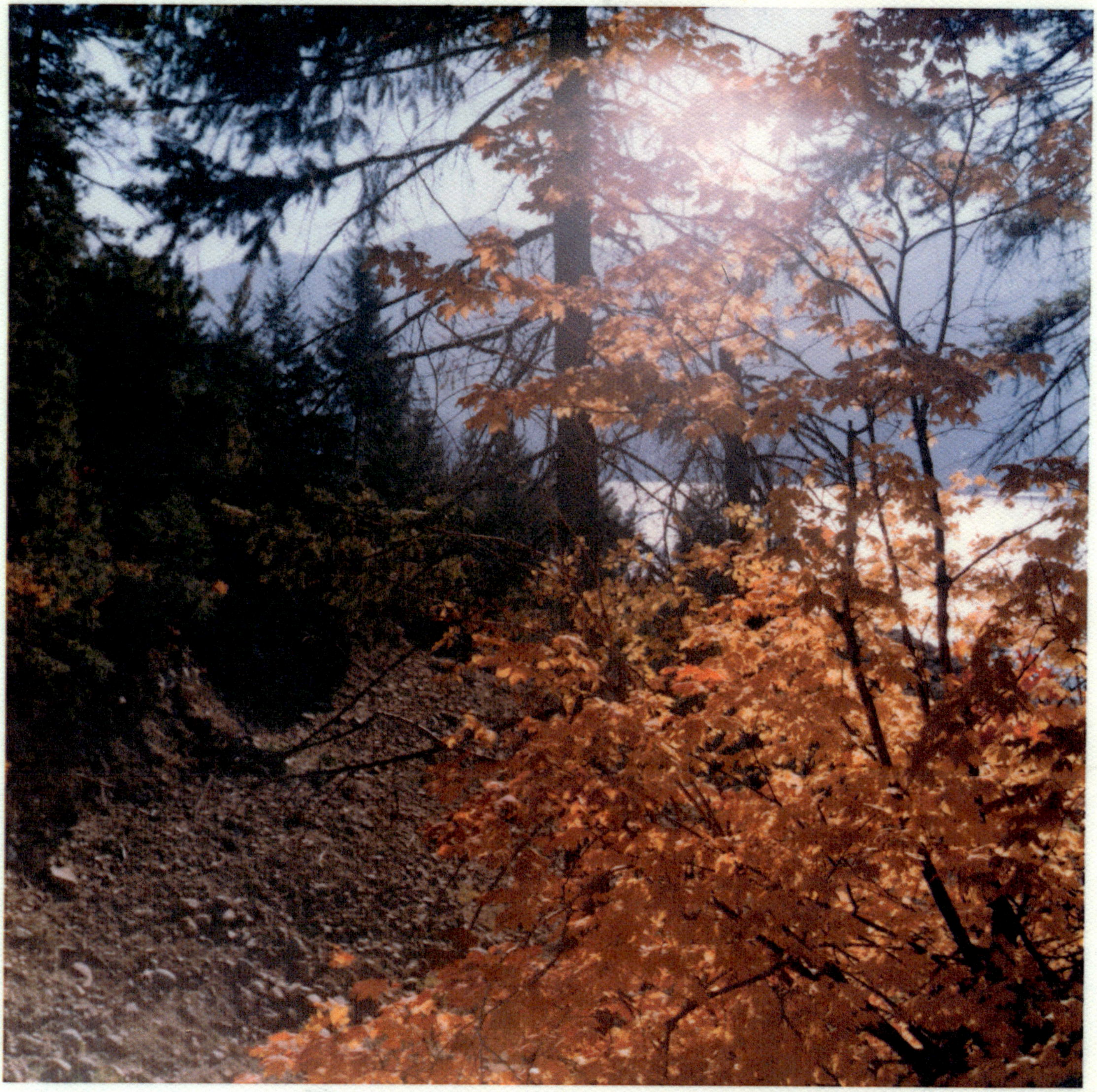

Kittitas Valley, about 1974, Grace Shaw Weber

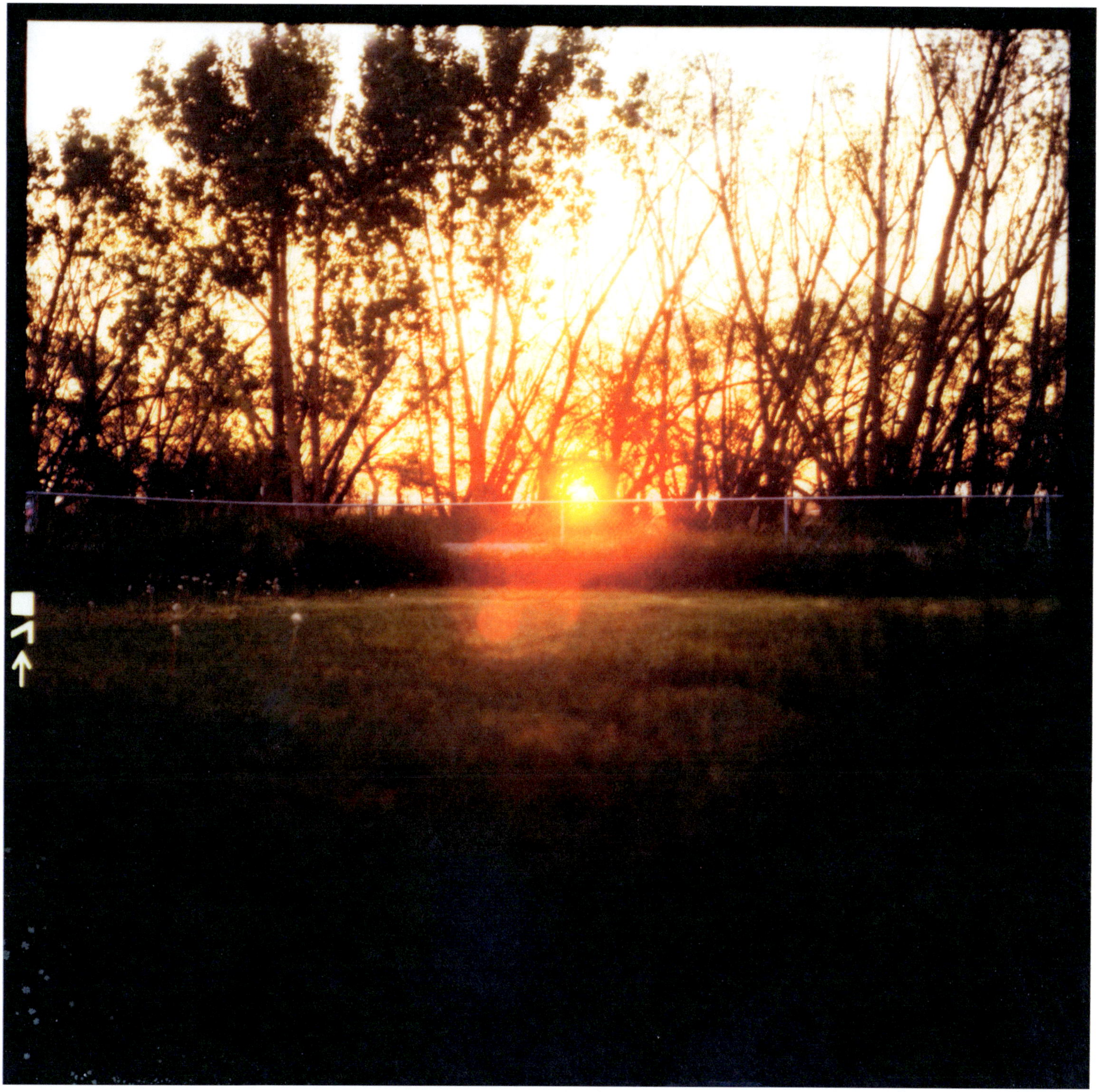

Quincy, mid 1950's, Grace Shaw Weber

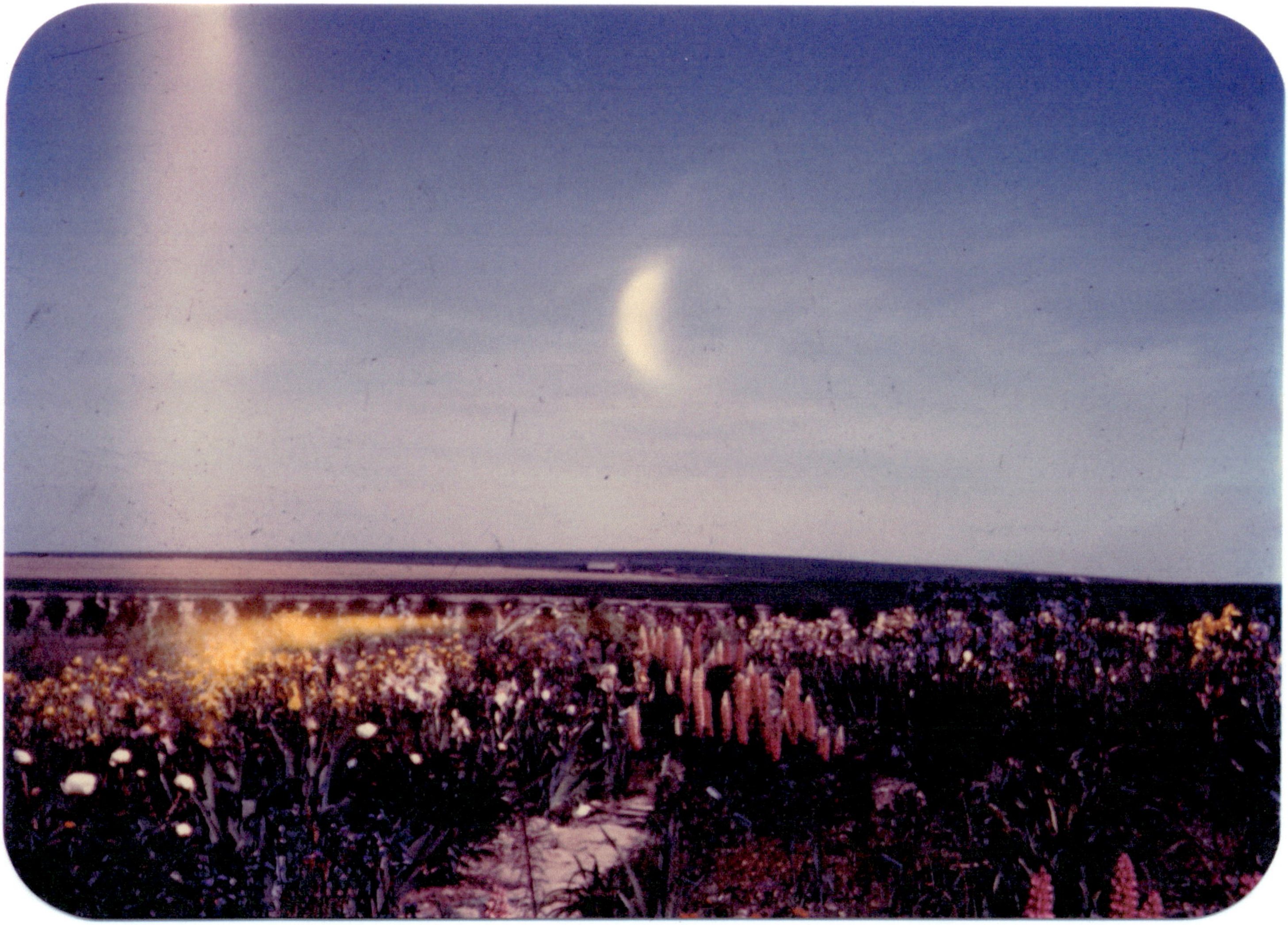

Kittitas, about 1976, Grace Shaw Weber

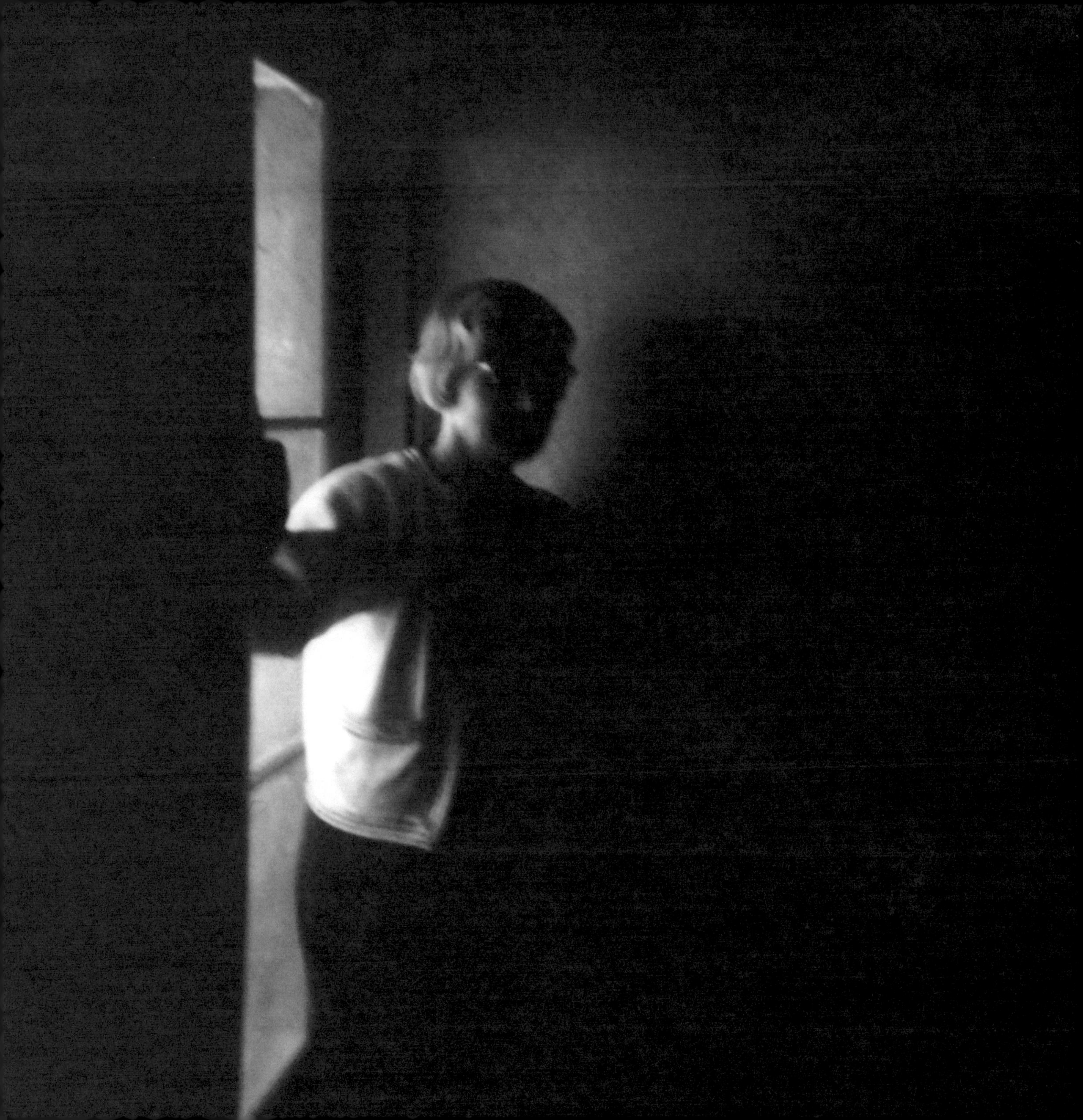

The Cartography of Farmers' Wives
Photography from 1915-1976

Colleen Weber Borst
Grace Shaw Weber
Emma Heckner Shaw

Cover photograph: Grace Weber Shaw
Layout and cover design by Mend My Dress Press

Published by Mend My Dress Press, Tacoma, WA
MendMyDress.com
MendMyDressPress@gmail.com

ISBN-10 0-9850131-2-5 (paperback)
ISBN-13 978-0-9850131-2-7 (paperback)

Manufactured in the United States of America

Many of the photographs in this collection have minimal digital enhancement. Century old fingerprints remain on negatives, and colour shifts on slides and prints have largely been preserved. I wanted the effects of time present on the photographs and film to be retained in these quiet meditations of a time and place that can not be revisited.

Made in the USA
Monee, IL
05 November 2026

67377158R00033